Francieli Kochem

The Insertion of Methodological Tools in Geography Teaching

Francieli Kochem

The Insertion of Methodological Tools in Geography Teaching

A Case Study in the State Schools/Colleges of Capanema, State of Paraná

ScienciaScripts

Imprint
Any brand names and product names mentioned in this book are subject to trademark, brand or patent protection and are trademarks or registered trademarks of their respective holders. The use of brand names, product names, common names, trade names, product descriptions etc. even without a particular marking in this work is in no way to be construed to mean that such names may be regarded as unrestricted in respect of trademark and brand protection legislation and could thus be used by anyone.

Cover image: www.ingimage.com

This book is a translation from the original published under ISBN 978-613-9-72261-7.

Publisher:
Sciencia Scripts
is a trademark of
Dodo Books Indian Ocean Ltd. and OmniScriptum S.R.L publishing group

120 High Road, East Finchley, London, N2 9ED, United Kingdom
Str. Armeneasca 28/1, office 1, Chisinau MD-2012, Republic of Moldova, Europe
Printed at: see last page
ISBN: 978-620-7-93782-0

SUMMARY

INTRODUCTION	**3**
CHAPTER 1	**6**
CHAPTER 2	**9**
CHAPTER 3	**12**
CHAPTER 4	**35**
CHAPTER 5	**52**

SUMMARY

The paper presents a brief chronology of the emergence of Geography as a science and as a discipline, emphasising some of its thinkers. It describes the relationship between Geography and Globalisation, and presents the results of field research carried out with Geography professionals working in State Schools in the city of Capanema, state of Paraná, Brazil. The research sought to identify the difficulties encountered by teachers when teaching their classes, as well as whether they already use different methodological tools, followed by a list of the tools mentioned by the teachers. Difficulties are becoming increasingly evident as society evolves, so innovation in teaching methods is necessary. In order to ratify the importance of using tools in Geography classes, bibliographical research was used, based on authors who have already written on the subject and who defend the use of cinema and TV, music and radio, newspapers and magazines, comics, ICTs, making maps and models, geographical games, oral reports and interviews, field classes and, finally, the textbook. The use of methodologies in geography lessons aims to provide a better assimilation of the content and contribute in a critical and reflective way to the formation of opinions.

Key words: Difficulties, geography teachers, methodologies.

INTRODUCTION

The lack of interest among students in Geography classes has long been discussed, a fact that causes concern among teachers in the state schools of the city of Capanema, in the state of Paraná. Most professionals no longer know how to instil an interest in the subject in their students.

In the context where I work, it can be seen that in addition to the students' lack of interest in the subject, there are several other difficulties that play a supporting role in the process and accentuate the difficulties in teaching classes. Whether it's indiscipline, the lack or absence of proper spaces for practical lessons, or the lack of resources to organise trips or even field lessons.

Specifically, Geography is considered by many to be one of the oldest sciences, the translation of the Greek word meaning the description of the Earth. Its emergence is related to observations made in geographical space, such as the direction of the wind, the phases of the moon, the direction in which rivers flow, in other words, natural phenomena.

To the extent that people have been able to organise their activities through these observations, countless scholars have dedicated their time to learning more about this science, which at first described the Earth, but then went on to study the relationships between society and geographical space, the latter being a product of social relations.

With the emergence of industry, Planet Earth began to experience a process called globalisation, which involves the whole of geographic space and society in a process of globalisation of information and situations. This process of globalisation also involves students from the municipality's state school network, both in urban and rural areas, encouraging them to interact in this process.

In this sense, teaching geography classes is becoming increasingly difficult because schools don't have as many technological materials and difficulties, and classes haven't met the expectations of students, who end up becoming disinterested in their studies.

Given this perspective and the local reality, the need arose to carry out an investigation and analyse the main difficulties that subject teachers encounter when teaching their classes in different schools/colleges in the municipality, as well as to point out the tools they already use to make their classes more interesting and creative.

Before describing the data collected, it was decided to describe the study area, which has been the target of many land disputes since the monarchical period, most notably in 1957 when squatters got together and led a revolt to expel the jagunços and their companies.

The city of Capanema was emancipated in 1951 and was colonised by German and Italian immigrants from Santa Catarina and Rio Grande do Sul. Today it is a quiet, small town in a frontier region whose economy is based on agriculture and commerce.

Before the Europeans colonised Capanema, Argentinians and caboclos already lived here, along with immigrants from various regions of Brazil (accentuated in 2014 by the construction of the Baixo Iguaçu Hydroelectric Power Station), and Asian and Latin American immigrants, Capanema is made up of a great diversity of cultures.

In terms of its objectives, this study is defined as exploratory research, because, according to Gil (2002, p. 41), "exploratory research aims to provide greater familiarity with the problem, with a view to making it explicit or building hypotheses, it involves a bibliographic survey", which provided support and backing for the problem in question, taking into account materials already published to consolidate information.

Data and information was collected through questionnaires administered to geography professionals in the municipality, with only two teachers opting not to take part. The data collected has therefore been transcribed into the text, some in tabular form and others in graphical form.

We chose to collect data from all the educational centres in the municipality's state school network, because the public who attend form a range with great variations and particularities, from socially and economically well-structured families to families with social and economic problems. Students who are concerned and interested in the quality of education, to students who are drug users and attend school out of obligation. In short, a public with a great diversity of culture and values, with whom the teachers work every day.

There are many difficulties in teaching the subject, which end up discouraging professionals from working with traditional teaching, based on summarising and memorising.

In the last chapters, the work describes some methodological tools that can be used in the construction of geographical knowledge, using authors who have already described the subject, to support and ratify the importance of using different methods in the transmission/construction of knowledge.

The aim here is not to pass judgement on teachers and their pedagogical practices, but to point out some useful and pedagogical tools, most of which are easy to acquire for working in geography classes and often go unnoticed.

Lima (2010, apud LOPES, 2010) points out that the teacher is the professional who, through their methodologies and existing resources or not, is able to articulate ideas and transmit knowledge, facilitating knowledge so that the student can understand it better. The use of a variety of tools in the subject's classes will encourage students to develop their own opinions, criticism and reflections and to be able to compete in a globalised society.

CHAPTER 1

The origin of geography

Considered by some to be one of the oldest academic disciplines, geography, a Greek word that epistemologically translates as "description of the earth" (SCHNEEBERGER and FARAGO, 2003, p.13), emerged in Greece. The Greeks dominated a large part of the western world and, in fact, when they were concerned with conquering new territories, they used everyday experiences, such as the direction of the winds, the phases of the moon, the direction in which the rains came, knowledge of the physical environment, in short, natural phenomena.

Over the centuries, many scholars have shown an interest in learning more about this science, which at first was considered to be just a description of the land, but as it has aroused human interest, it has evolved, according to numerous studies, observations and experiments, to become a comprehensive and interdisciplinary science. The contributions of Germans such as Humboldt, Ritter and Ratzel, and Frenchmen such as Bloch and Febvre, Braudel and La Blache, among others, have created concepts, many of which are used today (DINIZ FILHO, 2009).

Stefanello (2011, p. 24) points out that "until the Middle Ages, Geography was centred on studies related to astronomy and cartography and among its main functions were the organisation of routes and the survey of natural resources for exploration". Thus

Geography studied only the stars, maps and locations - this was the traditional current. It was only at the end of the 19th century that Geography began to study the relationship between society and nature, receiving many contributions from the naturalist and traveller Humboldt, the philosopher and historian Ritter and his successor Ratzel (STEFANELLO, 2011).

Still in line with the author, the post-war period saw the emergence of the new geography current, also known as theoretical-quantitative, which disregarded observation and replaced it with statistical data (STEFANELLO, 2011). This current of thought worked with statistical models that served to instigate economic growth, so ecological space was not mentioned.

It is worth noting that "The intense exploitation of nature, the increase in social inequality and a political movement of the popular classes fighting for a broad reform of society" (STEFANELLO, 2011, p. 28) fuelled the emergence of critical or radical geography in the 1970s. In this context, Milton Santos stands out for discussing the concept of space and instigating reflection on the organisation of space and its social structure.

The perception current in geography has been followed in Brazil since the 1970s, defending the thesis that "the spaces we find in our lived worlds are, above all, constructed spaces - made by man and consequently communicating human intentions and meanings" (RELPH, 1979, p.11). It therefore advocates all kinds of connections and relationships between man and the lived environments.

In the light of the above, geography has gone through numerous phases and evolved, according to Schneeberger and Farago (2003, p.13), "acquiring a comprehensive and complex meaning as it consolidates itself as one of the human sciences". It begins to be observed to the extent that the need for evolution, adaptation and knowledge arises, to the extent that human beings begin to question geographical phenomena and, through observations, experiences and interaction with the environment, obtain explanations.

Currently, geography's area of activity is wide-ranging, not just limited to physical aspects, but also human aspects, relating and studying the relationship between the two. Thus, there is physical geography, which is subdivided into several branches, focussing on physical aspects such as fauna, flora, hydrography, geomorphology, climatology and others (CERQUEIRA and FRANCISCO, on-line).

In human geography, the object of study is the human being acting on the physical environment and the relationships resulting from this action, such as anthropology, sociology, economic geography, political science and others

(CERQUEIRA and FRANCISCO, on-line). In this context, one cannot radically separate the areas that geography studies, as both are interconnected and go hand in hand, i.e. geography studies society acting on the physical environment.

CHAPTER 2

GEOGRAPHY AND GLOBALISATION

What is the relationship between geography and globalisation?

To answer this question, we need to know the origin of globalisation and its meanings.

Ianni (1994) describes that at the end of the 20th century and the beginning of the 21st century, the social sciences were faced with new objectives: to think of the world as a global society, with its geographical, economic, demographic, political, historical, cultural and social processes as a whole.

For Cotrim (2005, p. 530), globalisation consists of:

> A new process of reorganisation of economic productive forces in international terms has gained intensity. This new organisation of capitalism is based on a series of political and economic practices involving the public sector (states), private companies, universities and non-governmental organisations (NGOs).

In other words, Campos and Miranda (2005) describe globalisation as a show that can't stop, that must leave the world speechless, seducing the most miserable inhabitants of the planet with pyrotechnic and techno-science spectacles.

For Rattner (1995, p. 68):

> Globalisation began in the post-war period, with the accelerated and uninterrupted expansion of the internationalisation of the economy, shaped by the growth of trade and foreign investment, at rates much higher than the increase in world production (GDP - Gross World Product) [...].

The authors Alves and Oliveira (2013, p. 244) define globalisation as "the accelerated growth of economic integration between countries, resulting from the removal or reduction of barriers to the free flow of goods and capital around the globe". In this sense, globalisation is part of a process of increasing flows and goods.

It is clear that the definition of globalisation varies from author to author according to their vision. For Therborn (2001), globalisation received five central types of discourse at the end of the 1980s, the first being economic, then socio-political, socio-critical, cultural and social. This process encompasses a variety of issues in its discourses.

According to the aforementioned author, globalisation encompasses relations linked to the economy, buying, selling and exchanging goods and commodities. As for socio-political, it is linked to the role or function of the state and some kind of organisation. Socio-critical is a new way of seeing this whole process as an enemy of cultural values and social justice. Cultural because it presents globalisation as paths, flows of information and culture and, finally, social which presents concerns related to the environment (Therborn, 2001). Globalisation is understood to be a process of overvaluation that involves the whole world through economic, social, political, environmental and ideological processes.

Porto-Gonçalves (2012, p.12) states that "the overvaluation of the global scale reaches its peak through the affirmation of those who make use of this global scale: the large transnational corporations, multilateral organisations such as the World Bank, the IMF and the WTO". He also emphasises that environmental issues have never been addressed so much and, at the same time, huge areas of the planet have been devastated.

In the light of the above, it can be understood that globalisation has several definitions and they are related to the economy, politics, culture and the environment, because this process takes place in geographical space, in the relationships between societies.

With globalisation, the world has become a small place (even though it's a big one), since we can get wherever we want by means of transport, find out what's happening anywhere in the world at the exact moment it happens and communicate with any country in the world using these technologies. In short, with globalisation we connect with the whole world: from the local to the global.

Since the 1970s, geography has been interested in studying the relationships between society and physical space, while globalisation is a process inherent to geography, since without this science it would not be possible for globalisation to emerge, as it is a process of technological development connecting information networks, transport, trade and culture, shortening the world. This was only possible thanks to geographical studies.

We can therefore say that globalisation is a process that directly involves geography, because it is a process of technological development and interaction between society and space, which takes place in geographical space, the object of geography's study.

CHAPTER 3

TEACHING GEOGRAPHY IN STATE SCHOOLS IN PARANÁ, THE CHALLENGES OF TODAY

3.1 Location of the Study Area

To carry out the study, the area was delimited in order to analyse the characteristics of the population and the city, such as its location, a brief history of its foundation and its socio-economic characteristics. The city chosen for this study was Capanema, located in the south-west of the state of Paraná, in Brazil. This delimitation was made due to the author's residence being in this city, as well as the educational spaces in which she works. See location below:

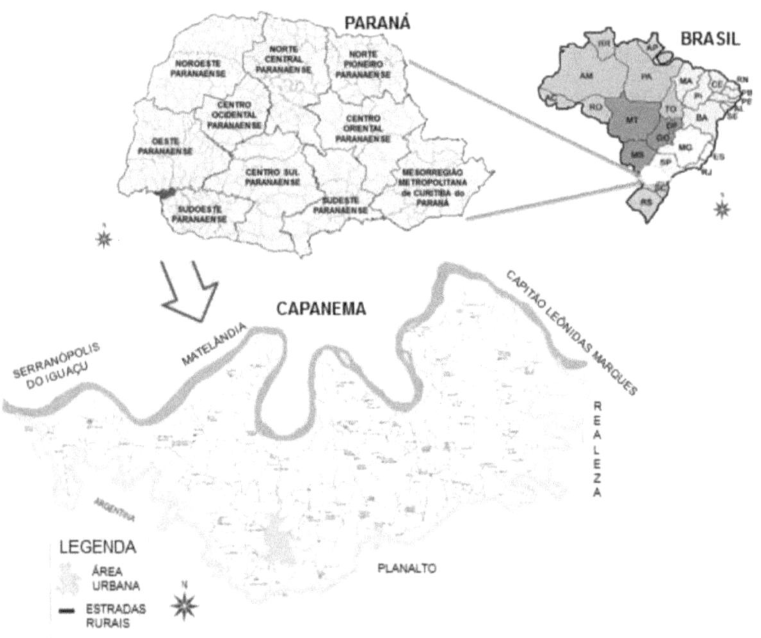

Fig.1 - Location map of the city of Capanema, state of Paraná, Brazil.

Source: Author's archive.

The state of Paraná was colonised through three occupation fronts: Paraná Tradicional, Frente Norte and Frente Sudoeste e Oeste (map 2). As Capanema is located in the south-western region of the state, we will only emphasise the south-western front.

Map 2 - Occupation fronts
Organised by the author.
Source: MACHADO, Brasil Pinheiro. Synopsis of the Regional History of Paraná. Curitiba, Historical and Geographical Institute of Paraná, 1951.

The colonisation and occupation of the lands that today belong to Capanema went through processes of disputes between Brazil, Argentina and Paraguay, as well as armed conflicts with hundreds of deaths (many of them unrecorded).

The first conflict took place between 1890 and 1895 and was called the **Question of Palmas** or **the Question of the Missions** (see map 2). In this conflict, Argentina claimed territories located in the west of the states of Paraná and Santa Catarina (DALLANORA, 2016).

Engineer Guilherme Schuch, better known as the Baron of Capanema, acted in this territorial dispute between the two countries, defending the border region that should belong to Brazil.

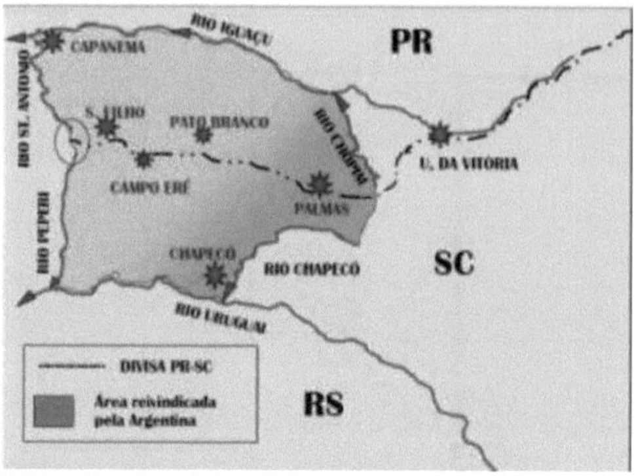

Map 3 - Region of the Palmas Dispute

Source: Norberto Boddy

Militeli (2016, p. 46) states that "in 1881, the Argentinians began exploiting yerba mate in Misiones and later in western Paraná, in large waves". This was only possible due to the presence of fertile land favourable for agricultural activities.

"Since the beginning of the frontier's advance, violence, conflicts and deaths have been part of the daily life of frontiersmen" (MYSKIW, 2009, p.18). The land conflicts began during the monarchy and lasted until 1957, when the settlers organised a revolt that put an end to the conflicts and oppression.

The occupation of the south-west of Paraná, with emphasis on:

> The fields of Palmas and Guarapuava were founded in the first half of the last century [...] as a result [...] of official measures, with the city of Guarapuava being founded by order of the government of the Province of São Paulo in 1819 and Palmas appearing towards the end of the century on the initiative of the government of Paraná (PADIS, 1981, p. 147-148).

According to Costa (2007), at the end of the 19th century, 18 population centres were created on the banks of various watercourses in areas close to Guarapuava, at the initiative of the government.

remained stagnant, sticking to subsistence activities. According to the author, when Paraná broke away from São Paulo in 1853, it was left with an extensive area of western Santa Catarina (COSTA, 2007).

Years later it resulted in an armed conflict between the caboclo population, inhabitants of the region who planted and raised animals, called pelados (for having their heads shaved due to lice), other authors refer to the pelados as posseiros (for having possession of the land without documents) and the peludos who represented Brazilian state and federal power and defended foreign companies such as Lumber and the Brazil Railway Company hired by the government to build a railway linking São Paulo to Rio Grande do Sul (CUNHA, 2016).

This conflict, which took place between 1912 and 1916, was called the Contestado War[1] and resulted in a barbarity without size, because while the pelados, led by the Italian monk João Maria, fought with sticks and stones, the peludos fought with guns and cannons (COSTA, 2007). The war ended in 1916 with the furry victors and part of the state of Paraná was incorporated into Santa Catarina. See map below.

Map 4: Contested War Region

Source: Cunha, 2016, online.

Another conflict that took place in the south-western region of Paraná was the Revolt of '57. Battisti (2006, p. 04) points out that "the context of the Revolt

[1] Contestado War My Paraná RPC. Available at: https://www. youtu be. com/watch?t=378&v=4082LGP3A8Y Accessed 21 Oct. 2018.
Contestado War 100 years. Available at: <https://www.youtube.com/watch?t=40&v=MW6_NTETUOA> Accessed on 21 Oct. 2018. Contestado - "War of God and the Devil". Available at: <https://www.youtube.com/watch?t=59&v=iPx- 7FqkyU> Accessed 21 Oct. 2018.

of 1957 was marked, at a national level, by new forms of capitalist exploitation in the countryside, manifesting itself in the various aspects that make up the Brazilian reality". In order to better understand the context of the Revolt, we need to start by describing national events that were linked to the revolt.

From 1930 onwards, through a policy of occupying Brazil's empty spaces, Getúlio Vargas tried to realise the strategy of the so-called "march to the west". According to Carneiro (1981, p. 55),

> [...] the march to the west was the first effective policy of the federal and state governments to integrate new areas into the process of expanding the economic frontier, the aim of which was to move the labour force to new fertile regions in order to start increasing food production for the country's booming urban centres.

This settlement process was reinforced with the creation on 13 September 1943 of the Federal Territory of Iguaçu through Decree-Law No. 5,812, which lasted until 1946. See map below.

Map 5: Federal Territory of Iguaçu

Source: Infographic, Gaveta do povo.[2]

After the creation of the Federal Territory of Iguaçu, huge tracts of land were created to facilitate control of Brazilian territory and encourage colonisation.

> **"Through decree number 300", he returned to the state's domain** huge tracts of land that stretched from the south-west to the division with Santa Catarina and which

[2] Available at: https://www.gazetadopovo.com.br/vida-e-cidadania/um-territorio-chamado-iguacu-bky41 lfjwksv2j8j851ds9etq/

> belonged to the Companhia Brasileira de Viação e Comércio (Braviaco), a subsidiary of the São Paulo - Rio Grande, which in turn was controlled by the Brazil Railway Co. (Wachovicz, 1985, p. 141).

Those plots of land that had not yet been allotted, such as Gleba das Missões, belonged to the states of the Federation. In 1920, Paraná turned it into payment for the construction of a railway branch linking Guarapuava to RFFSA (Rede Ferroviária Federal S/A). However, the company failed to fulfil its obligation, which resulted in a contentious dispute with the state of Paraná, which was trying to reclaim the Gleba das Missões (ZATTA and VANNINI, 2016).

Map 6: Location of Gleba das Missões

Source: MUNICIPALITY OF FRANCISCO BELTRÃO. Cultural Notebook - Memory Project. Department of Culture. Year 1, n°1 - Francisco Beltrão, 2007. Public Domain.

Cattelan and Castanha (2016) argue that in 1943 the federal government created CANGO (General Osório National Agricultural Colony) with the aim of speeding up the settlement of the region and the creation of the Federal Territory intended by the union, whose initial objective was to "benefit the large land grabbers in the south of the country" (WACHOWICZ, 1985, p. 180). The purpose of CANGO was to distribute land to settlers from Rio Grande do Sul and Santa Catarina, provide them with housing, agricultural tools, seeds, dental and medical care. In just a few years, this colonisation system attracted thousands of families who came to the region and made it one of the most developed.

The conflicts and land disputes escalated when José Rupp won a court case against Brazil Railway Compapanhy, which had owed him for 18 years for sleepers supplied for the construction of the São Paulo - Rio Grande railway.

After several attempts to reach an agreement, Rupp allied himself with Mario Fontana, a friend of Governor Lupion who exerted influence with the Federal Government, creating Clevelândia Industrial e Territorial Ltda. (CITLA) with the aim of colonising the Southwest, encompassing lands from the Glebas das Missões (including CANGO) as well as the headquarters of the districts of Francisco Beltrão, Santo Antônio and Capanema (BATTISTI, 2006).

Also, according to Battisti (2006), the state opposition denounced Rupp and the National Security Council forbade notaries from registering CITLA deeds, because "the Federal Constitution forbade the sale of land in border areas without the authorisation of the National Security Council" (VORPAGEL, 2006, p. 04). The federal government's aim was to protect the land in the border region, keeping it under its control to prevent foreign invasions and appropriation.

When Lupion returned to office in 1956, he forced Fontana to cede part of the Gleba das Missões to the companies "Comercial e Agrícola Paraná Ltda." and "Apucarana Ltda." In order to force the settlers to sign the debts, they used thugs to rob the settlers, beat them, kill them and rape them. Countless atrocities. Some groups of squatters began to organise themselves and set up ambushes to kill the jagunços. They had the help of the farrapos from Rio Grande do Sul, famous for their bravery and leadership. Among them were Pedro Santin, the Bello brothers and Robertinho, who led hundreds of settlers who rose up against the CITLA and APUCARANA companies in the border region (VORPAGEL, 2006, p.10).

> Santin's group took Capanema, the road was blocked, the APUCARANA company office was burnt down and the bandits fled to Santo Antônio. Around 2,000 armed settlers occupied the Capanema region. This was the place where the squatters first **"cleared the area", expelling the company and the jagunços. The situation in that** town was resolved with the direct intervention of State Police Chief Pinheiro Júnior who, in agreement with the settlers, managed to put an end to the conflicts in the border area. The solution for Santo Antônio was missing (VORPAGEL, 20016, p.11).

On 10 October 1957, thousands of squatters gathered and seized the main municipalities where the CITLA offices were based, evicting them and replacing the authorities.

The 1957 Posseiros uprising was victorious and its leaders became famous, but land ownership only came with GETSOP years later. Around 1950, GETSOP came into action with the aim of resolving the conflicts and legitimising the land, under the mandate of the President of the Republic Jânio Quadros, who declared the Gleba das Missões and part of the Gleba Chopim of public utility, titling more than 50,000 urban and rural properties.

The municipality of Capanema was dismembered from the municipality of Clevelândia by State Law No. 790 of 14 November 1951 and elevated to the category of municipality with definitive installation on 14 December 1952. Still in November 1952, the first election was held for the position of Municipal Mayor. Its first inhabitants were Argentinians and caboclos, who planted yerba mate and lived off subsistence farming. As the years went by and various conflicts unfolded, the Capanema region began to be populated by descendants of German and Italian origin from Rio Grande do Sul and Santa Catarina.

Little by little, human settlements sprang up, many of which, thanks to the fertility of the land, the dynamism of their creators and the opening of new communication and transport routes, made extraordinary progress. The pioneering explorer of the region was Octávio Francisco de Mattos, who after having held high public office in other municipalities, including that of Mayor of Clevelândia, decided to penetrate the interior of the hinterland and lay the foundations of a new settlement. Once the obstacles had been overcome, the intrepid bandeirante saw the emergence of a settlement on the edge of the border, which was given the name of Capanema.

The name of the municipality is a tribute to the eminent Brazilian geologist and geographer, Guilherme Schuch, the Baron of Capanema, who was the organiser and first director of telegraphs in Brazil (IBGE, 2007). The struggle and valour of the settlers for legal possession of the land is represented in the city's coat of arms, in which the eight towers represent the greatness of the city. The colour green symbolises hope, honour, green fields and the hope of a bountiful harvest. Silver symbolises peace, work and friendship, while black symbolises austerity, prudence and moderation. The gold and the pigs' heads recall the municipality's economic wealth, represented by pig farming and agriculture. Beans and soya are the main products of the land. And finally, the colour red symbolises the dedication and courage of the people of Capanema.

Fig.1: Coat of Arms of Capanema - Paraná

Source: Brasão, 2018.

According to data from the Brazilian Institute of Geography and Statistics (IBGE), the municipality of Capanema, located in the state of Paraná, has 18,526 inhabitants and a population density of 44 inhabitants per kilometre^2 . The municipality is located on the border with Argentina and borders other municipalities such as Planalto, Realeza, Capitão Leônidas Marques, Céu Azul, Matelândia and Serranópolis do Iguaçu.

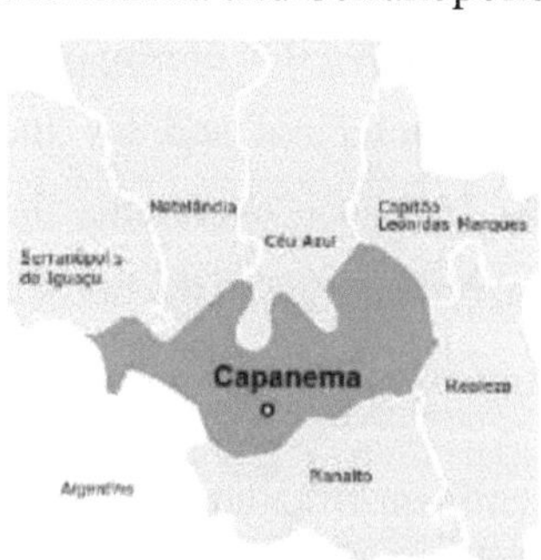

Map 7: Municipalities bordering Capenama - Pr

Source: IPARDES

NOTE: ITCG Cartographic Base (2010).

The relief of the city of Capanema is gentle, made up of slightly undulating plateaus, with small plains bordering the Iguaçu River. Among the mountains in the municipality, the following stand out: Serra Pelada, between the city centre and the district of São Luiz, and the Urutu, Pinheiro and Cedro hills.

Fig. 2: Geographical position of the city of Capanema-PR

GEOGRAPHICAL POSITION	INFORMATION
Altitude (metres)	368
Latitude	25^0 40'19"S
Longitude	53^0 48'32 " W

Source: Caderno Ipardes - 2018.

According to Ipardes (2018), the products that stand out in production are: Garlic, Peanuts (in husk), Rice (in husk), Potatoes, Sugarcane, Onions, Beans (in grain), Tobacco (in leaf), Cassava, Watermelons, Melons, Corn (in grain), Soya (in grain), Tomatoes and Wheat (in grain). It also produces a huge variety of fruit: Avocado, Banana (bunch), Persimmon, Yerba mate (green leaf), Fig, Guava, Lemon, Papaya, Mango, Pear, Peach, Tangerine and Grape.

Capanema's economy is based on raising cattle, sheep, horses, goats, pigs, buffaloes and chickens, and there is even a slaughterhouse in the town. From this information, it can be concluded that the municipality of Capanema is an inland municipality with its own characteristics. Its main economic activities are agriculture and industry.

The population is very diverse, predominantly of European descent, with a smaller number of Asian, Caboclo and Argentinian descendants. The municipality currently has a significant number of immigrants, labourers from various Brazilian states who have come to work on the construction of the Baixo Iguaçu Hydroelectric Power Station.

As far as the economy is concerned, there is a variation of social classes, with those who have the most purchasing power being traders and some settlers. However, the majority of the population belongs to the lower middle class.

The main conflicts that occur in the town include petty theft and drug trafficking, and rarely homicide and sexual abuse. In terms of quality of life, it can be considered a quiet, calm town, with reasonable distances to reach the hospital and the city centre. It's a great place to live peacefully.

3.2 Geography in the State Schools of the Municipality of Capanema

In order to analyse the current teaching of geography in state schools/colleges in the city of Capanema, state of Paraná, exploratory research was used, which aims to provide more familiarity with the subject in question, including bibliographic surveys (GIL, 2002), in order to directly observe the activities carried out by geography teachers, as well as to identify the factors that contribute to the use of different methodological tools in the teaching of geography. Other objectives make up the work, such as instigating students' interest in the subject and pointing out methodologies that can help in the acquisition of knowledge.

Systematic exploratory field observations were also used, which according to Silva and Menezes (2001, p. 21) seek to "describe the characteristics of a given population". The research was carried out at the Rocha Pombo, São Cristóvão and Padre Cirilo State Schools, the Santa Cruz State School, the Castelo Branco State School, the Duas Barras State School, the Pinheiro State School, the Antônio Francisco Lisboa State School and the Rui Barbosa State School, located in the city of Capanema, state of Paraná.

Map 8: Location of schools/colleges in the city of Capanema, Paraná.

Source: Author's archive.

In order to achieve the study's objective, a questionnaire was drawn up for the teachers, containing open and closed questions. With respect for the research subjects, a Free and Informed Consent Form was drawn up and delivered to each of the schools in the municipality (Appendix I). Afterwards, a pre-test was carried out to ensure the clarity and effectiveness of the instrument, where, through a quick conversation with the geography teachers who agreed to take part in the research, the main factors that hinder their teaching were listed.

Once the pre-test had been carried out, there was no need to make any changes to the questionnaire, so we started collecting information from the schools. Annex I contains the complete questionnaire for checking the tables, according to Silva (2009), which makes it easier to visualise when presenting the results.

The aim was to translate the information and data gathered through questionnaires with open and closed questions applied to the teachers into figures, which were analysed and classified. The research was qualitative because it considered observations and interpretations of phenomena and the attribution of meanings. In order to give greater credibility to the research, authors who have written about the difficulties of teaching classes were used, as well as authors who emphasise the use of different methodological tools for better acquisition of knowledge.

The information obtained through the open-ended questions was analysed using Bardin's (2009) technique, which requires a more in-depth interpretation of the object of research, making it important to read between the lines. According to Bardin (2009), content analysis is a set of research techniques which, through an objective, systematic and quantitative description of the manifest content of communications, aims to interpret these same communications.

Table 1 shows the schools surveyed and the number of teachers who agreed to take part in the research and completed the questionnaire.

Table 1: Number of schools surveyed and total number of teachers who answered the questionnaire.

Schools	N^0 of Teachers
Pigeon Rock	2
St Christopher's	2
Father Cirilo	1
Santa Cruz	2*
Castelo Branco	1
Two Bars	1*
Pinheiro	1*
Antônio Francisco Lisbon	1
Rui Barbosa	1
Total: 9 schools	8

Source: Information from the questionnaires
*Teachers who work in more than one school, already included in the total number of teachers.

Table 1 shows that questionnaires were administered in the nine state schools in Capanema, with 8 teachers taking part: 2 from the Rocha Pombo State School, 2 from the São Cristóvão State School, 1 from the Padre Cirilo State

School, 2 from the Santa Cruz State School, 1 from the Castelo Branco School, 1 from the Duas Barras School, 1 from the Pinheiro School, 2 from the Antônio Francisco Lisboa School and 1 from the Rui Barbosa School, totalling 8 geography teachers, 5 of whom work in two schools.Table 2 shows the identification characteristics of the teachers from the schools that took part in the study.

Table 2: Characteristics of School/College Teachers

Schools	Sex	Age	Length of professional career
Pigeon Rock	Female	61	43
	Female	27	2
St Christopher's	Female	58	38
	Female	61	29
Father Cirilo	Female	26	4
Santa Cruz	Male	29	8
	Female	26	4
Castelo Branco	Female	57	30
Two Bars	Female	57	30
Pinheiro	Female	57	39
Antonio Francisco Lisbon	Female	57	39
Rui Barbosa	Male	29	8

Source: Information from the questionnaires

Table 2 shows that the teachers who work at the Rocha Pombo State College are both female, one is 61 years old and has worked in education for 43 years, while the other is 27 years old with two years' experience in geography education.

The teachers at Colégio Estadual São Cristóvão are 58 and 61 years old respectively, with 38 and 29 years' experience in education.

The teacher from Colégio Estadual Padre Cirilo also works at Escola Estadual Santa Cruz, is 26 years old and has four years' experience in education. The teacher who works at Santa Cruz also works at Rui Barbosa, is 29 years old and has 8 years' experience.

The teacher who works at Castelo Branco and Duas Barras is 57 years old and has 30 years' experience in education.

The teacher, who works at Pinheiro and Antônio Francisco Lisboa, is 57 years old and has 39 years' experience in education.

Regarding the second question in the questionnaire about the use of different languages in the teaching of geography, all the teachers responded unanimously, even with different words. They all said that the use of different languages is necessary because our students are already in contact with technologies, and traditional teaching based on oral explanations and rote learning no longer stimulates or instigates the search for knowledge.The following table shows the responses of the research participants regarding the possibility of using different methodologies in their classes.

Table 3: Is it possible to use different methodologies in your classes?

Schools	YES	NO	MAYBE
Pigeon Rock	X X		
St Christopher's	X X		
Father Cirilo	X		
Santa Cruz	X X		
Castelo Branco			X
Two Bars			X
Pinheiro	X		
Antônio Francisco Lisboa	X		
Rui Barbosa	X		

Source: Information from the questionnaires

According to the table above, most of the teachers believe that it is possible to use different methodologies in geography classes, except for the teacher who works at the Duas Barras and Castelo Branco schools, who justified her answer "maybe" by the fact that the schools are in the countryside, difficult to access, with few students and most of the time do not have the materials needed to develop differentiated classes.The table below discusses the question about the difficulties encountered in teaching classes and the answers given by the teachers who took part in the research.

Table 4: Do you find it difficult to teach geography classes?

Schools	YES	NO	SOMETIMES
Pigeon Rock	X		X
St Christopher's	X X		
Father Cirilo			X
Santa Cruz	X X		
Castelo Branco	X		
Two Bars	X		
Pinheiro			X
Antônio Francisco. Lisbon			X
Rui Barbosa	X		

Source: Information from the questionnaires

From the data in the table above, it can be seen that the majority of teachers chose the option "yes", that they find it difficult to teach their classes. Only one teacher from Rocha Pombo, Padre Cirilo and Pinheiro, who also works at Antônio Lisboa, chose the option "sometimes".

The next table describes the (fifth) question regarding the difficulties encountered by professionals in the school environment. To make the answers easier to understand, the data will be broken down into numbers and percentages.

Table 5: Difficulties in teaching at Capanema/PR state schools/colleges - 2018.

DIFFICULTIES FACED IN TEACHING GEOGRAPHY	NUMBER OF TIMES CITED	%
Students uninterested in the subjects being taught	12	100%
Badly educated students	10	83%
Students without materials	9	75%
Little/lack of teaching materials	8	67%
Indiscipline in the classroom	8	67%

Laboratory Unavailable	8	67%
Few Hours Activities	7	58%
Lack of resources (tour/travel)	6	50%
Inadequate salary	4	33 %
Few lessons per week (little time per lesson)	3	25%
Little space in the school for different activities	3	25%
Students with frequent absences	2	16%
Lack of interest from other teachers in participating in projects	1	8%
Others	1	8%
Lack of textbooks	0	0%
Lack of pedagogical support / encouragement	0	0 %

Source: Information from the questionnaires

From the data collected, it can be seen that the item "Students uninterested in the subjects being taught" was chosen by all the teachers as the biggest problem, as the greatest difficulty they faced during their lessons, thus receiving 100 per cent of the vote.

The second item listed as the greatest difficulty encountered by teachers received 10 votes, with "Badly educated pupils" totalling 83%. Lack of education and, consequently, disrespect for the school's rules, demotivate professionals in their activities, in addition to the fact that teachers feel frightened.

The third item chosen with 75 per cent approval, receiving 9 votes respectively, was "Students without materials", which makes it difficult to carry out classroom activities as well as monitoring and teaching support.

The items "Little/Lack of teaching materials, Indiscipline in the classroom and Unavailable laboratory" received 8 votes simultaneously, totalling 67%. In view of this data, it can be understood that the difficulties in imparting knowledge are accentuated when there are unruly students, when there is a lack of teaching materials and the computer lab is often not working, there is no internet, or it is very slow, crashing computers.

The option "Few Hours of Activities" was chosen and received 7 votes, totalling 58%. "Lack of resources (excursions/trips)" received 6 votes, corresponding to 50% of the total. The lack of resources to organise trips and geographical excursions ends up discouraging professionals from getting students interested in geography, since contact with the object of study is of the utmost importance. Another issue that should be emphasised is that, due to the limited time available to prepare, correct and organise content and materials for lessons, teachers end up taking materials home and using their free time to carry out these activities.

"Inadequate salary", a low salary, was chosen by only 4 teachers who judged this item to be a difficulty, because it doesn't stimulate and motivate them enough to work in education these days in the face of ever-increasing difficulties. With 4 votes, it totalled 33%.

In seventh place, receiving 3 votes, totalling 25%, were "Few classes per week (little time per class) and Little space in the school to carry out different activities". According to some changes that have taken place in education, some classes have had their number of lessons reduced, especially sixth grade, encouraging teachers to organise themselves within the time available, a fact that is not always possible due to the sum of various difficulties that each teacher encounters in their classes.

Another item listed concerns the lack of space in the school to carry out different activities, such as a specific laboratory for geography to build models, a dedicated space with sound reduction to work on parodies, films, bingos and theatre.

Receiving just two votes was the item "Students with frequent absences", listed by teachers who work in rural schools due to the weather conditions and the financial difficulties of families who, at harvest time, give preference to helping their children with agricultural activities. This item accounted for 16 per cent.

Only 1 teacher chose the item "Other teachers' lack of interest in participating in projects". He pointed out that he finds it difficult to carry out projects at school due to his colleagues' lack of interest in taking part in interdisciplinary projects.

The option "other" reasons/difficulties encountered by teachers was selected by only 1 professional, accounting for 8% of the total, who described feeling a lack of support from parents, feeling that the management doesn't demand attention and monitoring from parents, emphasising that he misses the pedagogical team to support teaching practice, since the team is always solving problems of indiscipline and student absenteeism and, finally, emphasising the feeling of impotence and fear when having to teach students under the influence of narcotics.

As for the items "Lack of textbooks and Lack of pedagogical support/encouragement", these did not receive any votes, totalling 0% of the choices listed by teachers as difficulties in teaching geography classes. It can therefore be seen that there is neither a lack of textbooks nor a lack of pedagogical support and encouragement for geography lessons.

Question 6 asked whether the teacher interviewed had ever used different methodologies to explain the content, and the data is shown in the graph below.

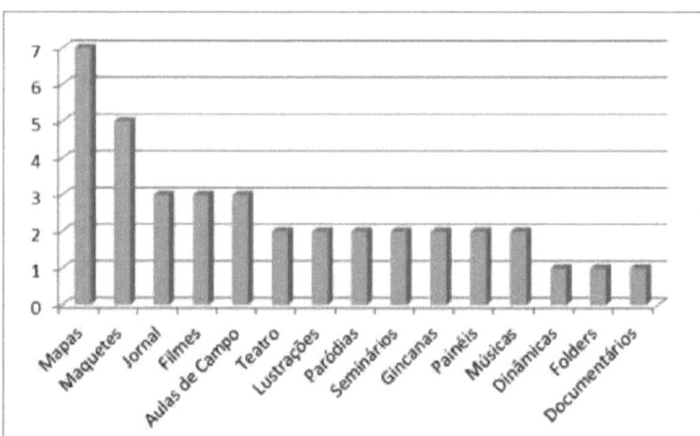

Graph 1: Have you worked with different methodologies to explain content? Which ones?

Source: Information from the questionnaires

According to the information collected, all the teachers who answered the questionnaire had already used a different methodological tool. The graph above shows the tools mentioned in descending order, with the most cited being maps, models, newspapers, films, field lessons, theatre, illustrations, parodies, seminars, gymkhanas, panels, music, dynamics, folders and documentaries. It can be seen that all the teachers have already looked for alternatives to make their lessons more exciting and dynamic, providing students with various situations and methods for acquiring knowledge.Question 7 asked whether the teacher had ever had to change their lesson plan due to difficulties they had faced; if they answered yes, they had to mention which difficulties had prevented them from carrying out what they had planned. As all the teachers ticked the "yes" box, the difficulties mentioned are shown in graph form to make them easier to understand.

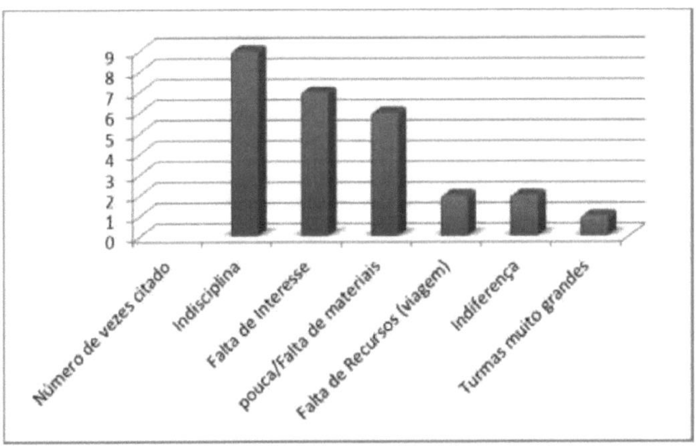

Graph 2: Difficulties responsible for the change in Lesson Plans
Source: Information from the questionnaires

As all the teachers replied that they had already had to change their lesson plans, the graph above shows the main difficulties listed by the teachers as being responsible for the changes in lesson plans, such as student indiscipline, lack of interest in the activity, limited availability or lack of teaching materials, lack of resources to carry out field activities, as well as trips and excursions.

Student indifference towards teachers and classes with up to 40 students were also cited. These factors, combined with other difficulties of time and space, end up discouraging teachers from innovating and looking for new teaching methods.

The answers to question number 8 will be organised in graph form, as this will illustrate the teachers' responses to the frequency of use of the methodologies described in the question.

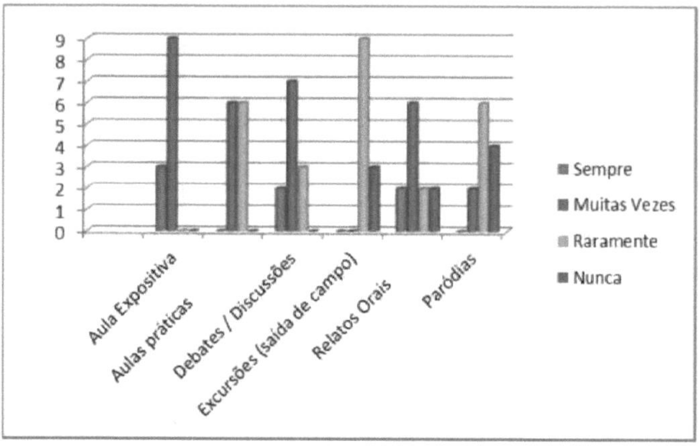

Graph 3: Frequency of use of methodologies
Source: Information from the questionnaires

Regarding the frequency of methodologies used, it can be seen that lectures are often used, followed by debates, practical lessons and oral reports. Rarely are activities such as excursions (field trips), practical lessons and parodies developed.Question 9 asked the teachers to indicate how often they used the methodological tools listed in a table. A graph was drawn up using the results.

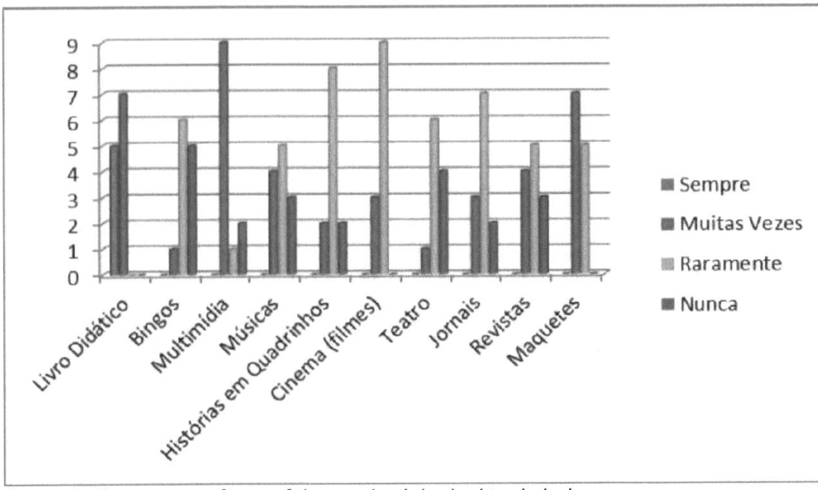

Graph 4: Frequency of use of the methodological tools below
Source: Information from the questionnaires

The use of multimedia, models and textbooks received many votes and were often listed as options. Rarely are films, comics, newspapers, theatres and geographical bingos used. From this information, it can be seen that other methodological tools are used in addition to the textbook, such as multimedia and the construction of models; however, the use of newspapers, comics, theatres, bingos, magazines and music are rarely used.

The rare use of these methodological tools is a reflection of the difficulties that teachers face in the school environment, because the tools are used when the class profile is favourable to their use and when materials and spaces are available to carry out the activities. Without this set of tools and (pedagogical) materials, combined with the availability of resources and spaces, it is very difficult for teachers to successfully fulfil their role, which, according to Milton Santos, the importance of education for the formation of the individual is indispensable:

> The real aim of education is not to arm the citizen for a war, that of competition with others. Its purpose, which is less and less sought after and less and less achieved, is to train people who are capable of situating themselves correctly in the world and influencing the improvement of human society as a whole (SANTOS, 1998, p. 126).

This can only be achieved when a teacher is able to use all the available methods and tools to encourage students to be critical and form ideas. It is important to emphasise that for a teacher to have comprehensive and problematising knowledge, and not stick to limited content, the insertion of methodological tools must be present in their practice.

It can be seen that there is often an intention to break away from traditionalism in the act of teaching, but the means and resources to achieve investigation and critical formation of ideas are outdated or lacking, making it difficult to do so. In this sense, the use of different methodologies in the classroom becomes difficult, but it should be emphasised that their use does not imply sporadic changes in lesson plans, just their adaptation to the materials available in the environment.Cordeiro and Oliveira (2011) point out that the search for new innovative methodological tools that stimulate the student's cognitive development is the most appropriate way to instigate students' interest in geographical knowledge to be developed in the classroom.

CHAPTER 4

Bases for the use of methodological tools in geography teaching

Through the information obtained from the research, we realised that there are many difficulties encountered by geography teachers working in the city's state schools/colleges. Not just problems related to the students, but to the whole school structure.

In this context, as a teacher, I believe that we should put these barriers and difficulties aside and try to work as best we can with what we have to help in education and the construction of knowledge. After all, our students are already involved with so many modern technological attractions that it doesn't make sense to teach only textbook content or always the same tools.

I would emphasise that the use of different methodologies in geography lessons allows students to build interdisciplinary, contextual knowledge, making it clear that geography is a science linked to life, and therefore to everyday life.

"Why innovate if the salary remains the same? If the government won't see the performance? If the students are always disinterested and the situation is likely to get worse?"

It's not our interest here to discuss the performance of teachers or their work, but to emphasise that educating is not a pastime, a means of earning a salary, or an activity that can be *"pushed around all your life and thought everything is fine!"* (emphasis added). Teachers have a duty to train and educate critical citizens, so that they are able and prepared to compete in the globalised world.

Using different methodologies to teach geography will not only make a difference for the teacher, but especially for the student, who will learn without having to experience traditional teaching. As for the salary, it will remain the same, but the greatest gain will be that of a conscious educator, an educator who is sure that he has done everything to improve the assimilation of learning, contributing to the critical and reflective formation of the student.

The teacher's mission is to encourage students to seek answers, not to teach them to memorise a book and accept what is described in it, the content ready and finished, but to encourage them to seek knowledge. The use of methodological tools in Geography classes will help students to expand their knowledge in an interdisciplinary and contextualised way, leading teachers to provide moments that instigate research (RUDNICK and SOUZA, 2010).

Another function of the teacher is to be the articulator of ideas, to transmit knowledge, to correlate and instrumentalise, to use the contents as they are, articulating them with the students' daily lives, in order to stimulate the different intelligences, leading them to form their own opinions (LIMA, 2010).

Didactic potential is not just about studying new verbal and non-verbal methodologies, it goes much further, expressing the interests and objectives of those who produce it. Therefore, when geography teachers prepare their lessons, they will do so with the aim of instigating their students by proposing questions about the object or problem being studied, making them seek answers, thereby expanding their knowledge and discernment.

It should be emphasised that in order to use methodological tools in geography lessons, you don't have to change the whole lesson plan, just adapt it and add some materials. Believe me, it's possible! This doesn't mean that all teachers will be able to work with countless different methodological tools in the face of the difficulties already described, but rather to try to convey geographical knowledge in a more creative way.

The following pages will describe some of the tools that can be used in the classroom, such as cinema and television, music and radio, newspapers, magazines and photographs, comic strips, information and communication technologies, maps, models, games, the use of oral reports and interviews, field lessons and textbooks.

4.1 The use of cinema and television

The use of cinema and television in the classroom contributes to the students' better development and understanding of the proposed subject. "Cinema gives us leisure and entertainment, but also access to information and scenarios at a low cost and quickly" (PEREIRA, 2013). Some forms of expression such as anguish, fear, repression and dreams can express the contexts of a particular time and/or society.

The choice of films, documentaries and cartoons can be made at the beginning of the school year, linked to the didactic content. In this context, the following films can be mentioned: "The android hunter, An Inconvenient Truth, The Last Hour, The Chain of Good, The Diary of a Motorcyclist". It's important to know some basic elements of the material, such as a brief summary, the era in which the story is set, some supporting texts such as interviews with the director and authors, reviews of the films published in the newspapers, and it's also good to know the tastes of the class in order to better accept the instrument. First, a questionnaire is drawn up, a script that expresses the intention behind the choice of film, documentary or drawing, which will guide the students in understanding and assimilating the content and the teacher's didactic intentions.

After the film, documentary or drawing and the questionnaire, the teacher can organise discussion groups, where, once formalised by the teacher, they can write down and illustrate the message that the material brings. Then there is a closing, a summary of the film, in which the teacher uses the opinions and approaches of the discussion groups, and correlates them to the curriculum content, making for a pleasant closing in order to understand that teaching geography means using different methodologies.

In order to use television in the classroom, it is of the utmost importance that there is also planning, both in the choice of programmes and content, making it clear that once the tools have been used, there will be a job to be done, not a laser.

In the course of these activities, doubts arise as to whether the teacher necessarily needs to be a researcher, have training in the humanities that involves research, or whether they can adapt it to their teaching. In this context, Napolitano

(2002) advocates that being a primary and secondary school teacher means having intellectual autonomy, being able to plan and organise new techniques, new projects and using new tools to teach. Therefore, it is understood that it is possible to teach geography without depending on or using only and exclusively the textbook.

First, themes and material are selected: miniseries, soap operas, newspapers, both with cross-cutting and/or behavioural themes, but which are appropriate for the age groups. After accessing the material, the teacher guides the formation of discussion groups, pointing out some complementary perspectives that, in line with Costa (2011, p. 74), should point out:

> The establishment of content and the development of skills at the start of the activity; the articulation with themes and content suggested in the sequence of the course itself; the opening up to related and complementary themes, which can motivate the student to seek more knowledge on their own, based on the skills and content established.

At the end of the activity, the teacher closes by explaining the ideas and intentions initially planned, relating the results and discussions.

Stefanello (2011, p. 137) points out that "the universe of art is extremely rich for working with geographical content, as the methodologies range from the visual arts to the sciences and the plastic arts." The art of music, if properly planned and introduced, can lead students to a diversity of knowledge. The fact that it is part of everyday life makes its use more enjoyable and relaxing. It is important to select lyrics and repertoires according to the content. In this way, we can relate the parody, through the song interpreted, to its context of appearance, make a lyric with rhymes, contextualising it with the proposed content, making the students reflect critically, but constructively on the theme (AIRTON, 2012).

Its use in the classroom contributes to better student development and understanding of the proposed subject, because through the scenarios, images and environment the student will feel involved and assimilate more easily.

Other tools such as dance, drama and theatre, arts that involve bodily expressions, can be used, because through clothing we can analyse religious and/or social beliefs, and the stage set can take us back to a period in history. Through expression, ideas, feelings, symbols, psychological aspects and space can be analysed. In addition to these tools, the arts that denote psychic diction, such as cartoons, poems, comics and photos, can be analysed as part of an observational analysis (AIRTON, 2012).

4.2 The use of music and radio in the classroom

Considered one of the oldest communication activities, music was used for communication even before speech. The sounds of birds, animals and various instruments and materials were propagated.

The use of music in the classroom, provided it is properly planned and inserted, can lead students to a variety of knowledge. The fact that it is part of everyday life will make it more enjoyable and relaxed. It is important to select lyrics and repertoires (not forgetting to combine or approximate the rhythms of the music with the tastes of the class), to research the author, the lyrics, the context in which it was written and whether it has become popular.

There are many song lyrics that can help with Geography lessons by describing situations, places, events, environmental problems, water wastage, in short, a diversity of themes and content. The authors Tortoreli and Paixão (2018) emphasise the lyrics of the following songs as a source of information and knowledge for Geography. They highlight the song "Planeta Azul" by Chitãozinho and Xororó, "Sobradinho" by Sá and Guarabyra, "Planeta Água" by Guilherme Arantes and "O Sal da Terra" by Beto Guedes.

Teaching Geography with music will become a different way of getting to know and learn Geography without having to memorise dates and books. According to Costa (2011), some historical events and the music used for this understanding and relationship are exemplified: - when teaching the study of Brazilian Indians and their participation in the formation of Brazilian society, you can listen to: - Jaboti Indians from Rondônia. Tchori, Tchori. Interpreter: Marlui, Miranda. In: MIRANDA, M. **Ihu:** todos os sons [S.I.]: Pau- Brasil, 1998. Track 01. - Juruna Indians from Mato Grosso. Ju Paraná. Performer: Marlui Miranda. In: MIRANDA, M. Ihu: Todos os sons. [S.I]: Pau- Brasil, 1995, track 4. These songs talk about indigenous culture, customs, religions and traditions, as well as making us reflect on the different realities of indigenous tribes, which, being from the north of Brazil, are generally little known by most Brazilians.

After the song has been played, the teacher is instructed to provide the lyrics and make them available to the students, who can analyse the lyrics, write new lyrics in groups, with melodies and rhythms, and organise a plenary session.

Just like other teaching tools, when radio is taken into the classroom it needs to be planned. After all, because it is an instrument used in people's daily lives, its power and influence can be turned into school knowledge as long as the various programmes it broadcasts are accompanied by activities and reflections guided by the teacher.

Planning involves choosing the content, selecting periods and events, natural disasters, problems. After choosing and distributing the topics, the teacher needs to obtain these materials from programmes and other radio productions by contacting the radio station.

We recommend contacting radios that have a wide reach and are old, radios that keep relevant material in their database, which can be documents, news, texts, recordings of speeches or interviews related to the subject, without forgetting to cite the sources. For example, CÂMARA radio[3] can be searched by entering the topic to be searched in the SEARCH field.

The teacher can propose a script to guide the students' research:

- ask the students to do some research on the period or event to be studied; listen to and analyse the material received from the radio; instruct them to take notes, then gather them in groups to exchange information, ask them to compare the material resulting from the research with the material received from the radio,

[3] Available at: <http://www.camara.gov.br/intemet/radiocamara/default.asp>. Accessed on 07 Nov. 2018.

ask them to present the results of their notes in plenary (COSTA, 2011).

The importance of guiding the student in synthesising the content seen using this method is highlighted. After all, this article proposes teaching Geography using different methodologies in order to make lessons more comprehensible, dynamic and that result in learning, valuing the critical formation of opinions.

4.3 The use of newspapers, magazines and photographs in the classroom

These are examples of other methodologies that can be used to complement geography lessons. The printed newspaper has some peculiar features which, when used coherently, can contribute to the teaching of the subject in the classroom. The fact that it is a daily newspaper portrays and accompanies everyday situations linked to events in "the country, the state, the region, the city or certain characters in highlights that no other means of communication records in such detail" (COSTA, 2011, p.123).

As it is printed material, it can be used long after the events have taken place and can be found, above all, in libraries that store the collections or in the programme offices of the newspapers themselves or in radio station databases.

It's important to emphasise to the teachers that even though newspapers present discourses of freedom, they have financial objectives, in other words, to sell news, sell adverts and make a profit. To the detriment of this issue, their printed news is the result of the objectives and political submissions of their advertisers, bringing subjects that arouse curiosity and interest to the reading public in order to make a profit.

After reading, reflecting on and analysing the material, the teacher selects materials linked to the content, makes them available to the students, guides them to compare the content of the newspaper with that of the textbook, proposes a plenary session to present the results and discussions, drawing a conclusion in the sense of comparing both materials.

Another example of using the material in the classroom is organising a Reporteca. The work can be done in groups and the students will have to research news items from newspapers or magazines, current and highly relevant topics. They can use posters or put together slides for the presentation, in which they will cut out the pictures or images and produce a text about the news item in their own words and present it to the class. The news stories presented in the Reporteca could be on a weekly or fortnightly basis, during which time various pieces of information will be made available for the groups to choose from.

The fact that the newspaper is an important means of communication and can be accessed daily by a large part of the population (in this case the newspaper presented by television stations, which can later be accessed via the internet or in print by subscription), provides a wide variety of news and information, which, in addition to students interacting with the current world, can develop their critical, conscious and reflective side with the help of the teacher.

Another type of printed material that is useful for supplementing studies are magazines. Their content is wide-ranging, from news to local, regional and international reports. When it comes to magazines as a methodological complement in geography teaching, we highlight geographic magazines, which contain accessible language and are made up of articles, dossiers, reports, interviews, information on national and international geography congresses and events, photos, illustrations and much more attractive information.

It should be emphasised that the research should be carried out on reliable websites and that the work should be subjected to rigorous proof-reading by masters, doctors and PHDs, that it should contain scientific articles, summaries, varied information on the most diverse geographical contexts, current problems faced by teachers, methodological and didactic options.

You can see the wealth of possibilities for its use in the classroom, in fact, useful tools with pleasing results only with the guidance of the teacher. Thus, after the teacher has selected the content and magazines, the material is made available to the students, so that they can read and analyse the material following a script provided by the supervisor. The debate on the material is held in plenary, with the activities carried out by the supervisor being finalised and concluded.

Printed magazines and newspapers have some peculiar features, because most of the time they are daily and record detailed facts, which, when used consistently, can contribute to a better understanding and acquisition of the content by the students.

Another useful tool in Geography lessons is the use of photographs, because due to the processes of transformation and change that occur daily in the geographical space, photos represent faithful documents of the reality recorded, and can be used in Geography lessons as a tool for transmitting knowledge and instigating critical reflection.

4.4 The use of comics in the classroom

The use of comics should be seen as a way of interacting with geography content, not just as an adversary to teaching. Just like other methods, the tool must be planned, the choice of theme must follow the curriculum content, with an emphasis on its application, especially in the age group, which is appropriate according to the intellectual development of the students.

This tool helps to complement lessons as long as it is planned, as mentioned above, with creativity and good judgement. Its use can be linked to content such as urban environmental problems, natural disasters, the consequences of anthropisation, among others.

It is important to emphasise to students when drawing up comics that they should contain objects that are characteristic of the periods or events such as clothing, means of transport, symbols, expressions on people's faces characterising obedience, submission, happiness, buildings with their models, shapes and materials with which houses, temples, churches and buildings were characterised in the period, time, used for its design stars such as the Sun, Moon, stars, or even clocks.

Some elements should serve as a basis to guide the students' work. The teacher could also suggest that the students carry out research in order to get to know some of the elements that characterise comics in time and space, without leaving them with the impression that it is just a pastime.

4.5 The use of new information and communication technologies in the classroom

When using information and communication technologies (ICT) in the classroom, the computer, the internet and multimedia stand out, all of which are extremely useful due to the wide range of information they provide. These tools are currently gaining prominence in the lives of the people who make up education and are present as daily resources for communication, leisure and work (SALES, 2013).

The use of computers connected to the internet provides us with a wealth of information ranging from the Ancient Age to the Post-Modern Age, offering photos, drawings, figures, poems, messages, videos, information, articles, testimonies, books, studies, interviews, old documents, virtual tours, radio frequencies, music, films, games, newspapers, dictionaries, training courses, further training, Virtual Learning Environments, distance learning degrees, postgraduate courses, among countless others.

Using virtual programmes, you can obtain images from satellites positioned in different parts of the planet. The use of this programme is useful when studying environmental problems caused by anthropogenic action that are visible to the human eye, such as deforestation, fires, mineral extraction activities, desertification, among others.

The use of these technologies tends to broaden the possibilities for better understanding of the content, as well as sparking interest in their use in the field of research. In this sense, Sales (2013, p. 02) advocates that:

> There is no way that schools can remain oblivious to this reality and not include in their daily teaching routine the use of technological resources that enable the enrichment of educational activities and the growth of teachers and students, as well as the expanded use of new languages.

When using different methods in Geography lessons, it's worth emphasising the importance of planning the use of the resource, as well as the stipulated time and some reliable electronic addresses, because while the technological medium can be of great value, it can also be useless when not correctly guided and directed.

4.6 Map making

Teaching students to make maps encourages them to develop their skills and stimulates the knowledge they already have, because "cartography encompasses elements of science and art at the same time" (TORTORELI and PAIXÃO, 2018, p.166). According to the authors, maps have been used since prehistoric times to demarcate hunting and fishing territories (TORTORELI and PAIXÃO, 2018, p.167).

Over the years, the purpose of using and making maps has changed, until we reach the 21st century with an immense variety of objectives and elements arranged on maps.

Most of the time, the use of maps in geography lessons is no more than a mere copy, perfect by the way, and memorisation of types of vegetation, delimitation of territories, names of countries and capitals, among others. In this context, Passini and Almeida (2004, p. 19) advocate:

> [...] the most commonly used ways of working with cartographic language at school is through situations in which students have to colour in maps, copy them, write the names of rivers or towns, memorise the information represented on them. But this doesn't guarantee that they will build up the necessary knowledge to read maps and represent geographical space.

In light of this, it is important to rethink the way we work with maps in the classroom: do we still demand and live by a system of memorisation, or do we encourage the creation of maps according to each individual's ability?

It's worth pointing out that before encouraging students to create maps, it's important that they learn to read them, otherwise encouraging them to do so will be pointless. Therefore, use the title of the map to situate the student in the information that the map will cover, followed by reading and interpreting the legend. Finally, reading and observing the scales, which can be graphic or numerical, and the position of the cardinal points in order to situate the student in space. The use of mental maps is also important because, as well as stimulating the student to reason and remember the space to be drawn, it instils the knowledge already acquired to draw the symbols.

Another important tool for understanding space is sketching. This is a simple activity that doesn't require any technological devices or a lot of knowledge and mastery, just observation of the space to be drawn. This activity stimulates observation, reasoning, the development of motor skills, knowledge and a sense of direction and space.

Because Geography studies society and geographical space, it is of the utmost importance that Geography teachers include the making of maps and sketches in their lessons. These tools don't require the use of technology or travelling to represent the space on paper; a simple stroll around the schoolyard or the court surrounding the school is enough to carry out these activities.

4.7 Mock-ups

A very dynamic and useful representation of geographical space is the construction of mock-ups because, as well as encompassing physical aspects of the environment to be portrayed, they also stimulate creativity and make students recall their acquired knowledge. Considered an important support resource, the construction of models also encourages logical thinking and collective exchanges of knowledge. In this vein, the author describes its importance:

> [...] by working with information in relief, allowing a three-dimensional view of space, it brings the abstract closer to the real and, at the same time, makes it possible

> to build the concepts needed to understand two-dimensional representation (GOMES, 2005, p. 209, online).

The process of making a model involves many steps and instils curiosity in students. Firstly, it is necessary to observe the space to be built (a very important action when studying geography), then plan and organise the materials for construction and, finally, take action and build the desired space.

There are countless contents that can be represented using models, and it's up to the teacher to delimit them and coordinate the whole process, helping students to build a new, smaller-scale space.

4.8 Games in Geography Lessons

The theory of establishing a playful education is not something new, since ancient Greece and the Middle Ages emphasised the importance of playful activities in the teaching-learning process.

> Games allow freedom of action, inner drive, naturalness, attitude and, consequently, pleasure, which are rarely found in other school activities and should therefore be studied by educators as another pedagogical alternative at the service of integral development [...] (NEGRINE, 2011, p. 69 and 70).

It's no different in Geography lessons, where the use of game-related activities also contributes to the development of learning, but in a more dynamic way.

> [...] the introduction of play in Geography lessons allows students to assimilate content in a fun way, without the momentary worry of being assessed. Playful activities arouse students' interest in lessons, because when they make the object to be worked on, through handling and visualisation, they can relate the place where they live to Geography (SAWCZUKI; MOURA, 2012, p. 03, online).

By working with playful activities, students are encouraged to develop cognitive skills such as attention, observation, interaction and

participation, as well as revising and consolidating content, also helping to integrate students.

Examples of geographical games include memory games with the Brazilian states drawn on a certain number of tiles and their respective names on another. Dominoes games with the names of the countries and capitals of the American continent.

Bingos with the capitals of Brazil, Europe, territorial boundaries, climates, industries, Brazilian biomes and their economic activities, and the biomes of the world. In this game, the students will have to put together their own card by choosing names related to the theme, for example, if the theme is biomes of the world, the student will have to write some names of the biomes on their card and the teacher will have to come up with questions about the biomes, so that the answers are biomes: Savannah, prairies...

Another example of games are geographical trails. The teacher will have to provide the trail and the data, then prepare the questions with their respective answers on a separate piece of material. The use of gymkhanas with the capitals or on other content stimulates students' reasoning and helps them to retain their knowledge.

In view of the above, the use of games in Geography makes a significant contribution to revising and fixing content, as well as helping with social interaction and, through the way it works, contributing to the acquisition of geographical knowledge in a dynamic and freer way.

4.9 Oral accounts or interviews

The collection of interviews, testimonies and the fact that we can ask questions of people from the past is a very rich source for studying the past.

It takes us back to a more direct past or reality than reading texts.

Schmidt and Cainelli (2004) emphasise that oral history can be presented through oral life stories, which allow the interviewer to relate the interviewee's facts to the group in which they are inserted.

Not all objects of study in Geography have published, printed or photographed materials for research. This happens, for example, when we study the city's rivers.

Given this fact, oral accounts and interviews play a fundamental role in providing information about the object and giving researchers knowledge of the past, so that we can then carry out analyses and comparisons. There are many objects of study in the area that we can get to know better through oral accounts and interviews. Some even involve family members who keep private memories and photos of moments from the past.

> We can therefore say that memory is a constituent element of the sense of identity, both individual and collective, in that it is also an extremely important factor in a person's or group's sense of continuity and coherence in their reconstruction of themselves (POLLAK, 1992, p. 204).

With this in mind, there's no reason why Geography shouldn't use these sources to enrich its knowledge of the space in which it operates. Expanding knowledge, valuing space and understanding the activities carried out in this geographical area, as well as broadening students' knowledge, will encourage them to seek out and value these sources more.

4.10 Field lessons

The use of field classes in the study of Geography helps students to understand the relationship between theory and practice, because at the same time as it allows contact with reality, it awakens and motivates interest in the subject.

Stefanello (2011) emphasises that in field classes, students come into direct contact with their object of analysis, broadening their creativity and reasoning, and their capacity for construction.

Because it is a geographical space, the object of study of Geography, field classes contribute better to an understanding of space, place and the world, "linking theory to practice, through observation and analysis of the lived and conceived space" (LIMA E ASSIS, apud FALCÃO e PEREIRA, 2012, p. 112). Contact with real space stimulates students' thinking and enables them to better relate theory and practice.

For some authors, field lessons are practices in which students can see geography, see reality, see space, not just read and develop mental images and leave it at that.

> [...] the field lesson is an out-of-classroom/out-of-school activity that concomitantly involves school, scientific (or not) and social content with the spatial modality; social reality and its complex material and immaterial amalgam of traditions/new things. It is a movement that tends to elucidate sensations of strangeness, identity, ugliness, beauty, feeling and even rebellion in what is observed, interviewed, photographed and travelled (OLIVEIRA and ASSIS, 2009, p.196).

In this sense, field classes allow students to come into contact with reality, with the object of study, being able to analyse it, observe it, compare it and understand it on the basis of the theoretical content already covered.

It should be emphasised that contact with the object of study expands the students' imagination and provides more eloquent and enjoyable learning.

4.11 The use of textbooks in the classroom

Most teachers use this didactic resource without prior preparation, mechanically following the reading or explanation of the content, often without looking at figures, graphs, explanatory texts or reflective texts at the end of the chapter. Preparatory reading on the part of the teacher is necessary to better contextualise the subject, and they can look for or access other materials in order to complement the subject's lessons.

This article suggests that, as well as choosing a good textbook with accessible language and content that is enjoyable and appropriate for the age of the students, the teacher should be a geographer, attending national and international seminars, conferences, workshops, conference proceedings and subscribing to reliable, high-quality journals to improve and develop geographical knowledge.

After all, Geography is not a static human discipline/science; it is constantly changing, always depending on human actions or the weather. Therefore, it doesn't make sense for geography teachers to only use the textbook as a single, monotonous piece of material; they need to know how to use it and incorporate more content, news and information into it.

It can be seen that the textbooks made available by the government in the municipality's schools/colleges and that reach the educational spaces were written approximately 2 and 4 years ago, which justifies the importance of teachers preparing their lessons, organising the content and adding more information to it. It should be emphasised that the textbook is an important material that facilitates the teaching-learning process. However, when it comes to teaching, it is necessary for the teacher to get to know it and incorporate more information and update the data available in it.

In this context, it is understood that the textbook should be a complement to the classes, and the use of other methodologies should work alongside this material, so that one complements and supports the other, in order to contribute to the student's education, so that they don't feel trapped by the content and theme of the book, but that they know how to correlate them with real situations and form their own opinion.

CHAPTER 5

FINAL CONSIDERATIONS

After a brief chronology of the emergence of Geography as a science and discipline, emphasising some influential thinkers, the work pointed out the relationship between Geography and Globalisation, both of which act together in the geographical space. The former is a science that studies the relationship between society and nature, and the latter is known as a process of spreading information and knowledge to most places on Planet Earth in real time.

The second part of the work sought to locate and present the area in which the research was carried out, using maps to make it easier to understand. It also described the city's colonisation process, which was marked by many conflicts and at different times, and was made a municipality in 1951. It also located all the schools in the municipality, both in the urban and rural areas.

In order to make it easier to understand the local population, statistical and economic data on the activities carried out by Capanemenses were transcribed. From this, it emerged that the municipality is inland, located on the border, and is characterised by being a quiet municipality with good expectations and quality of life.

With regard to the objective of analysing the difficulties that teachers encounter when teaching geography, the data collected through the questionnaires was extremely important in identifying the difficulties that professionals encounter on a daily basis. Among the most frequently listed was student indiscipline, in second place was a lack of interest in lessons, followed by a lack of materials in schools to support lessons, a lack of resources to organise trips and field lessons, and in last place were students' indifference to teachers and classes with too many students.

Based on this information, the professionals were asked how often they used methodological tools in class, and they chose between the options always, often, rarely and never. The majority of votes for the often option went to lectures, followed by debates and discussions.

There are many difficulties in teaching geography classes, ranging from problems of indiscipline to a lack of resources to buy materials and/or go travelling. However, it was observed that most teachers, even in the face of the difficulties encountered, try to make their lessons creative and interesting.

The last part of the book mentions various methodological tools that can be used in Geography classes with the aim of getting students more interested in the subject. After studying the literature, the tools were listed one by one, describing examples of how they can be used, some even mentioning content. These tools have already been identified by authors as effective in provoking interest and enjoyment in lessons, and can be used in schools/colleges in the city of Capanema.

It's not in our interest here to pass judgement on the classes of the teachers surveyed, but to offer them ideas for effective, simple teaching materials, in the case of magazines, to stimulate students' interest in the subject, encouraging them to develop critical thinking.

REFERENCES

AIRTON, José. The Pedagogy of Enchantment: Working with Music in the Classroom. Available at http://www.construimoticias.com.br/asp/materia.asp?id=538. Accessed 04 Nov. 2018.

ALVES, Alexandre; OLIVEIRA, Letícia Fagundes de. Connections with History. 2ª ed. São Paulo: Moderna, 2013.

BARDIN, L. Content Analysis. Lisbon, Portugal; Edições 70, LDA, 2009.

BATTISTI, Elir. THE LAND DISPUTES IN THE SOUTHEAST OF PARANÁ: the land conflicts of the 1950s and 1980s. CAMPO- TERRITÓRIO: revista de geografia agrária, v. 1, n. 2, p. 65-91, aug. 2006. Available at: < www.seer.ufu.br/index.php/campoterritorio/article/download/11785/8291 Accessed 5 Nov. 2018.

BIESEK , Elizete Lírio; GANDRA, Edgar A. TERRITÓRIO FEDERAL DO IGUAÇU. Available at: <http://www.diaadiaeducacao.pr.gov.br/portals/pde/arquivos/724-4.pdf> Accessed 30 Oct. 2018.

CAMPOS, Flavio de; MIRANDA, Renan Garcia. The Writing of History: secondary education. Vol. Único. 1ª ed. São Paulo: Escala Educacional, 2005.

CARNEIRO, Maria Esperança. The Peasant Revolt of Formoso and Trombas. Goiânia, 1981, p.6 and 7. Agricultural Social Research Exchange Project. Agreement between CPDA/EIAP/FGV/ FORD Foundation.

CATTELAN, Carla; CASTANHA, André Paulo. The General Osório National Agricultural Colony (Cango) and the schooling process in south-west Paraná: 1948 - 1957. Oficina do Historiador, Porto Alegre, EDIPUCRS, v. 9, n. 1, jan./jun. 2016. Available at <http://revistaseletronicas.pucrs.br/ojs/index.php/oficinadohistoriador/article/viewFile/18680/14549> Accessed on 10 Nov. 2018.

CERQUEIRA and FRANCISCO, W. Physical Geography. Available at: <https://mundoeducacao.bol.uol.com.br/geografia/geografia-fisica.htm> Accessed on 15 Nov. 2018.

CORDEIRO, Joel Maciel Pereira; OLIVEIRA, Aldo Gonçalves de. A Aula de Campo em Geografia e suas Contribuições Para o Processo de Ensino-Aprendizagem na Escola. (undergraduate article). Revista Geografia (Londrina) v. 20. N.2, p. 099-114, May/Aug. 2011. Available at: < http://www.uel.br/revistas/uel/index.php/geografia/article/view/7416> Accessed on 11 Nov. 2018.

COSTA, Armando João Dalla. Teaching History and its Languages. Curitiba: IBPEX, 2011 (History and Geography Teaching Methodology collection; v. 7).

COSTA, Rosana Ruivo da. PARANÁ - THE OCCUPATION OF SPACE AND THE MECHANISMS OF LAND APPROPRIATION AS ELEMENTS THAT GENERATE AGRARIAN CONFLICTS. 2007. Available at: <http://www.gestaoescolar.diaadia.pr.gov.br/arquivos/File/producoes pde/artigo rosana ruivo.pdf> Accessed 7 Nov. 2018.

COTRIM, Gilberto. Global History - Brazil and General. Vol. Único. 8th Ed. São Paulo: Saraiva, 2005.

CUNHA, Carolina. Contestado War. 2016. Available at: <https://medium.com/@ Contestado/especial-100-anos-do-fim-da-guerra-do-contestado-9f1 e0c2fec97> Accessed 01 No. 2018.

DALLANORA, Cistina. FRONTIER MOVEMENTS: LAND CONFLICT AND COMMERCIAL TRANSACTIONS IN THE REALISATION OF PROPERTY IN THE RIO DO PEIXE VALLEY. XY Regional Meeting of History. 100 years of the Contestado War, Hydrography, Collections and Sources. Available at: < http://www.encontro2016.pr.anpuh.org/resources/anais/45/1468516877 ARQUIVO TextoanpuhCristinaDallanora.pdf> Accessed 9 Nov. 2018.

DINIZ FILHO, Luis Lopes. Epistemological Foundations of Geography. Curitiba: IBPEX, 2009 (History and Geography Teaching Methodology collection, v.6).

FALCÃO, Wagner Scopel; PEREIRA, Thiago Barcelos. The field lesson in the student's critical/citizen education: an alternative for teaching geography. Available at http://www.agb.org.br/XENPEG/artigos/GT/GT4/tc4%20(2).pdf. Accessed on 27 Oct. 2018.

GIL, Antônio Carlos. How to prepare research projects. 4ª Ed. São Paulo: Atal S/A, 2002.

GOMES, Iria Zanoni. 1957 Revolt of the Squatters. 3 ed. Cutitiba: Criar Edições, 2005.

IANNI, Octavio. Globalisation: New paradigm for the social sciences. Revista Estudos Avançados. vol.8 no.21 São Paulo May/Aug. 1994. Available at: <http://dx.doi.org/10.1590/S0103-40141994000200009> Accessed on 27 March 2018.

BRAZILIAN INSTITUTE OF GEOGRAPHY AND STATISTICS. Capanema, Paraná , Brazil. Available at : <https://cidades.ibge.gov.br/brasil/pr/capanema/panorama> Accessed 2 Nov. 2018.

PARANAENSE INSTITUTE FOR ECONOMIC AND SOCIAL DEVELOPMENT. CAPANEMA MUNICIPALITY STATISTICAL NOTEBOOK.
November/2018. Available at: <http://www.ipardes.gov.br/cadernos/MontaCadPdf1.php?Municipio=85760> Accessed on 14 Nov. 2018.

LOPES, Jaime Sérgio Frajuca. Teacher-researcher in Geographical Education. Curitiba: Ibpex, 2010 - (collection Metodologia do Ensino de Historia e Geografia;.v.4).

MILITELI, EDSON MATIAS. Representation of the Argentinian on the border(s). STRICTO SENSU POSTGRADUATE PROGRAMME IN SOCIETY, CULTURE AND BORDERS - MASTER'S AND DOCTORAL LEVEL AREA OF CONCENTRATION: SOCIETY, CULTURE AND BORDERS, 2016.
Available at: http://tede.unioeste.br/bitstream/tede/2939/5/Edson Militelli 2016.pdf Accessed 15 Nov. 2018.

MYSKIW, Antônio Marcos. The frontier as a travelling destination: The military colony of Foz do Iguaçu (1888-1907). Niterói: UFF, 2009 (Doctoral thesis).

NAPOLITANO, M. Como usar a televisão na sala de aula. 4. ed. São Paulo: Contexto, 2002ª .

NEGRINE, Airton. Learning and child development: symbolism and play. Porto Alegre : Prodil, 1994. v. 1.

OLIVEIRA, Christian Dennys Monteiro de: ASSIS, Raimundo Jucier Sousa de. Travessias em campo na Geografia escolar: a necessidade convertida para além da fábula. Educação e Pesquisa, São Paulo, v.35, n.1, p.195-209. Available at: http://www.scielo.br Accessed on 29 Oct. 2018.

PADIS, Pedro Calil. Formation of a Peripheral Economy: The case of Paraná. São Paulo : Hucitec. 1981. Co-edited in Curitiba : Secretariat of Culture and Sport of the Government of the State of Paraná. Economics and Planning: Theses and Research Series.

PASSINI, Elsa Yasuko; ALMEIDA, Rosângela Doin. Geographical space: teaching and representation. 13 ed. São Paulo, Contexto, 2004.

PEREIRA, Luiz Antônio de S. Films, documentaries and drawings and the teaching of geography. 10th NATIONAL MEETING ON GEOGRAPHY TEACHING PRACTICE, Porto Alegre, 2009. Available at <http://www.agb.org.br/XENPEG/artigos/GT/GT5/tc5%20(86).pdf> Accessed 23 October 2018.

POLLACK, M. Memory and social identity. Estudos Históricos: Teoria e História, Rio de Janeiro, vol.5, n.10, 1992.

PORTO-GONÇALVES, Carlos Walter. The globalisation of nature and the nature of globalisation. Rio de Janeiro: Brazilian Civilisation, 2012.

Rattner, Henrique. Globalisation: heading in one world? Rev. Estudos Avançados. vol.9 no.25 São Paulo Sept./Dec. 1995. Available at:

<http://www.scielo.br/pdf/ea/v9n25/v9n25a05.pdf> Accessed 25 February 2018.

RELPH, E. The Phenomenological Basis of Geography. Geografia, Rio Claro, v.4, n.7, p.1-25, Apr. 1979.

RUDNICK, Rosane; SOUZA, Sandra de. Teaching Geography and its Languages. Curitiba: Ibpex, 2010 (History and geography teaching methodology collection).

SALES, Mary Valda de Souza. Technologies, teaching and the training of geography teachers: new languages in the transformation of practice. 10th NATIONAL MEETING ON GEOGRAPHY TEACHING PRACTICE, Porto Alegre, 2009. Available at <http://www.agb.org.br/XENPEG/artigos/GT/GT5/tc5%20(29).pdf> Accessed on 29 October 2018.

SANTOS, Milton. O espaço do cidadão. 4 ed. São Paulo: Nobel, 1998.

SAWCZUK, Márcia Inês Lorenzet; MOURA, Jeani Delgado Paschoal. PEDAGOGICAL GAMES FOR TEACHING GEOGRAPHY (Article PDE, 2012). Available at: <http://www.diaadiaeducacao.pr.gov.br/portals/cadernospde/pdebusca/producoes pde/2012/2012 uel geo article marcia ines lorenzet sawczuk.pdf> Accessed 11 Nov. 2018.

SCHMIDT, M. A.; CAINELLI, M. Teaching History. Sao Paulo: Scipione, 2004 (Thought and Action in Teaching).

SCHNEEBERGER, Carlos Alberto; FARAGO, Luiz Antonio. Compact Minimanual of GENERAL GEOGRAPHY-theory and practice. 1. ed. São Paulo: RIDEEL, 2003.

SILVA, Ana Beatriz B. Restless Minds: ADHD: inattention, hyperactivity and impulsivity. Rio de Janeiro: Objetiva, 2009.

SILVA, Edna Lúcia da. MENEZES, Estera Muszkat. M. Metodologia da Pesquisa e Elaboração de Dissertação. Florianópolis. UFSC Distance Learning Laboratory, 2001.

STEFANELLO, Ana Clarissa. Didactics and Assessment of Learning in Geography Teaching. 2ed. Revised and updated. Curitiba: Ibpex, 2011. - (History and Geography Teaching Methodology collection. v.2).

THERBORN, GORAN. Globalisation and inequality: questions of conceptualisation and clarification. Sociologias, Porto Alegre, ano 3, n° 6, jul/dez 2001, p. 122-169. Available at: <http://www.scielo.br/pdf/soc/n6/a07n6.pdf> Accessed 29 Oct. 2018.

TORTORELI, Adélia Cristina; PIXÃO, Priscila Campiolo Manesco. Geography Teaching Methodology. Maringá -PR, 2017. Reprinted in 2018.

VORPAGEL, Edvino Knasel. The squatters' revolt in south-west Paraná in 1957. Available at: <http://www.gestaoescolar.diaadia.pr.gov.br/arquivos/File/producoes pde/md edvino knasel vorpagel.pdf> Accessed 21 Oct. 2018.

WACHOWICZ, Ruy Cristovam. Southwest Paraná: Occupation and Colonisation. Curitiba: Lítero-Técnica, 1985.

ZATTA, Ronaldo; Vannini, Ismael Antônio. Southwest Paraná and the military presence on the border: aspects of an "imagined community". Vozes, Pretérito & Devir Year III, Vol. V, N° I (2016). Available at: <http://revistavozes.uespi.br/ojs/index.php/revistavozes/article/view/104/134> Accessed on 15 Nov. 2018.

ANNEX I

Dear Teacher

You are being invited to take part in this research, the aim of which is to investigate the difficulties encountered by Geography teachers in the primary, *secondary* and *high school classroom in the state schools of Capanema - PR.*

You are free to refuse to take part, but by taking part you will be collaborating with the scientific community on the subject in question.

You will be taking part in the data collection and will express your opinion on the main difficulties, methods and methodological instruments in the questions listed below by means of a questionnaire.

All the information collected in this study is strictly confidential. Only the researcher will be privy to the data.

We count on your valuable participation!

QUESTIONNAIRE

1- Identification:

Name (initials only):

Sex: ( ) male ( ) female.

Age: ____

Year(s) in which you work: ______________________________

Subject in which you work: ______________________________

2)What do you think about using different languages in geography teaching?

____

____

____

3) Is it possible to introduce different methodologies during the course of the lessons?

( ) Yes ( ) No ( ) Maybe.

4) Do you find it difficult to teach your geography classes?

( ) Yes ( ) No ( ) Sometimes

ONLY ANSWER THE FOLLOWING QUESTION IF YOU TICKED YES

FOR QUESTION 4.

5) WHAT DIFFICULTIES?
Tick the option(s) below

( ) Students uninterested in the subjects being taught
( ) Badly educated pupils
( ) Students without materials
( ) Students with frequent absences
( ) Indiscipline in the classroom
( ) Laboratory Unavailable
( ) Few lessons per week (little time per lesson)
( ) Other teachers' lack of interest in participating in projects
( ) Little space in the school for different activities
( ) Little/Lack of teaching materials
( ) Lack of resources (tour/trip)
( ) Lack of pedagogical support / encouragement
( ) Inadequate salary
( ) Lack of textbooks
( ) Few Hours Activities

( ) Other, which? ______________________________________

6) Have you ever worked with different methodologies to explain content? ( )

Yes ( ) No ( ) I don't remember.

If yes, what were the methodologies?

7) Have you ever had to change your lesson plan/methodology because of difficulties you faced in the classroom?

( ) Yes ( ) No ( ) I don't remember

If you answered yes, write down the difficulty you encountered.

8) How often do you use the methodologies below:

Lectures	□ always	□ Often	□Rarely	□never
Practical lessons	□always	□Many times	□Rarely	□never
Debates / Discussions	□always	□Many times	□Rarely	□never
Excursions (field trip;	□ always	□ Often	□Rarely	□ never
Oral reports	□always	□Many times	□Rare mind	□never
Parodies	□always	□Many times	□Rarely	□never
0ther Qual ?				

9) How often do you use the methodological tools below?

Textbook	□Always	□Many times	□Rarely	□Never
Bingos	□Always	□Many times	□Rarely	□Never
Multimedia	□always	□Many times	□Rarely	□Never
Music	□Always	□Many times	□Rare mind	□Never
Stories in Comics	□Always	□Many times	□Rare mind	□Never
Cinema (Films)	□Always	□Many times	□Rare mind	□Never
Theatre	□Always	□Many times	□Rare mind	□Never
Newspapers	□always	□Many times	□Rarely	□Never
Magazines	□Always	□Many times	□Rarely	□Never
Mock-ups	□Always	□Many times	□Rarely	□Never
Map construction	□Always	□Many times	□Rarely	□Never
Sketches	□always	□Many times	□Rarely	□Never
Other. Which ones?				

Please use the space below for any suggestions you may have.

Thank you!

MIX
Papier aus verantwortungsvollen Quellen
Paper from responsible sources
FSC® C105338

Printed by Books on Demand GmbH, Norderstedt / Germany